AF545143

Elephants

WONDER STARTERS

Elephants

Pictures by ISOBEL BEARD

Published by WONDER BOOKS
A Division of Grosset & Dunlap, Inc.
51 Madison Avenue New York, N.Y. 10010

About Wonder Starters

Wonder Starters are vocabulary controlled information books for young children. More than ninety per cent of the words in the text will be in the reading vocabulary of the vast majority of young readers. Word and sentence length have also been carefully controlled.

Key new words associated with the topic of each book are repeated with picture explanations in the Starters dictionary at the end. The dictionary can also be used as an index for teaching children to look things up.

Teachers and experts have been consulted on the content and accuracy of the books.

Published in the United States by Wonder Books, a Division of Grosset & Dunlap, Inc.

Library of Congress Catalog Card Number 74-2755
ISBN: 0-448-09677-3 (Trade Edition)
ISBN: 0-448-06397-2 (Library Edition)

FIRST PRINTING 1974

Printed and bound in the United States.

This is a herd of elephants.
These elephants live in Africa.

This elephant is
the leader of the herd.
He is a big elephant.
He has big tusks.

This is a mother elephant.
She is giving milk to her baby.

Baby elephants are hairy.
They have no tusks yet.

This little elephant is five years old.
His tusks are growing.
He plays with the other elephants.

The elephants see danger.
They stand in a ring.
The baby elephants stay
inside the ring.

These elephants are frightened.
They run very fast.
They break down trees.
They are stampeding.

The leader of this herd
is very old.
A young elephant fights him.

Now the young elephant is leader.
The old elephant goes away.

This elephant lives in Asia.
It is a tame elephant.
Men often ride tame elephants
in Asia.

This elephant is working.
A man rides on his back.
The man taps the elephant's ears.
This tells the elephant what to do.

Some elephants work in sawmills.
They pick up logs
with their trunks.

An elephant's trunk
can pick up small things.
It can even pick up a pin.

Elephants often stand in water.
The water keeps them cool.
They fill their trunks with water.
They blow the water out again.

This elephant is having a bath.
It is lying down
in the water.
Some elephants are swimming.

Most zoos have elephants.
Once they were kept in cages.
Now they have more room.
They can move about.

These elephants are in a circus.
They can do tricks.

Once men hunted elephants.
They killed the elephants.
They took their tusks.

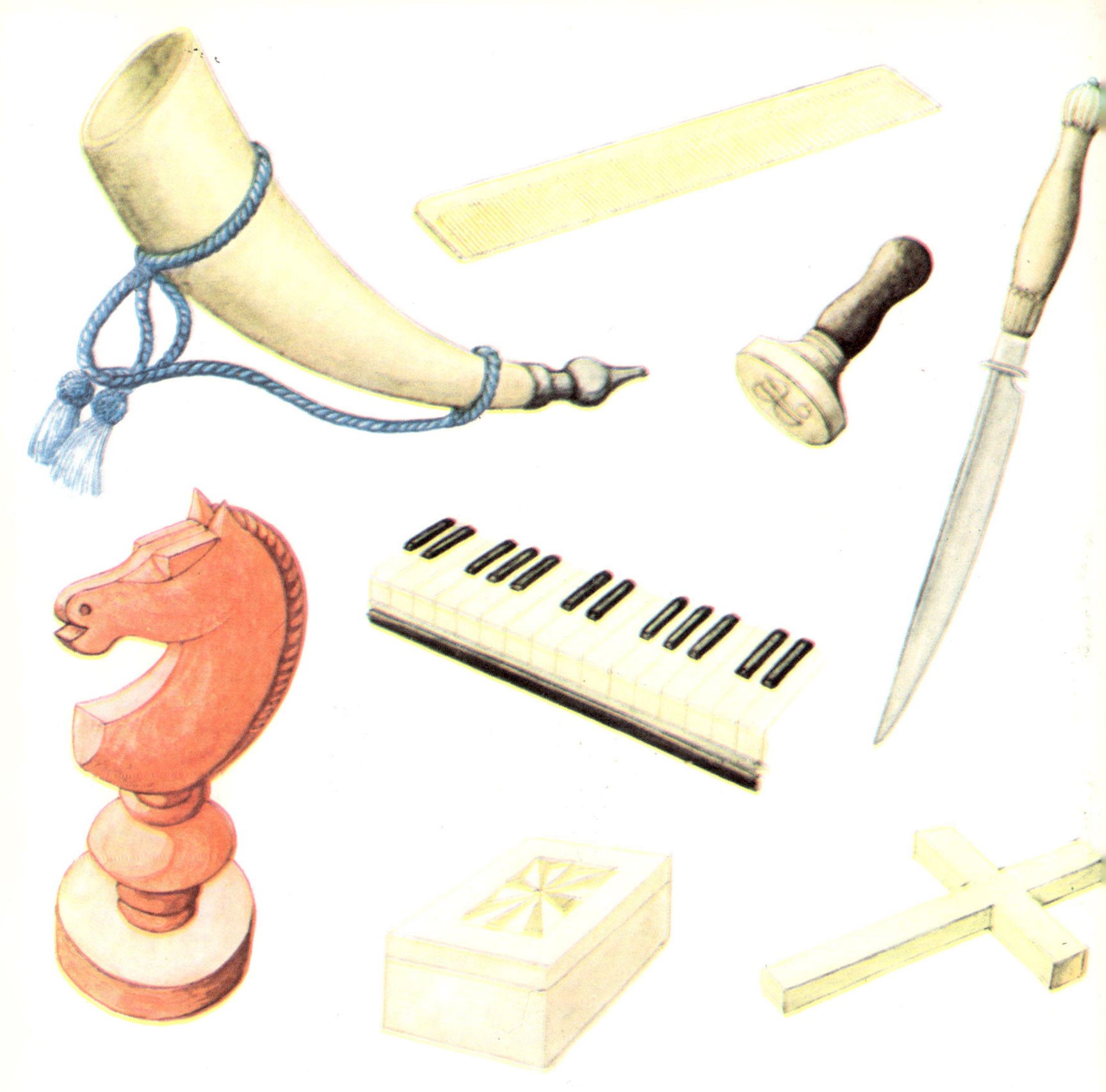

Elephants' tusks are really teeth.
They are made of ivory.
People use ivory
to make many things.

This man had forty elephants.
His soldiers rode on elephants.

These soldiers rode on elephants, too.
They lived in India long ago.
They went to war
riding on elephants.

This kind of elephant
lived three million years ago.
It was a hairy mammoth.
There are no mammoths left now.

Starter's **Elephants** words

herd
(page 1)

Africa
(page 1)

leader
(page 2)

tusks
(page 2)

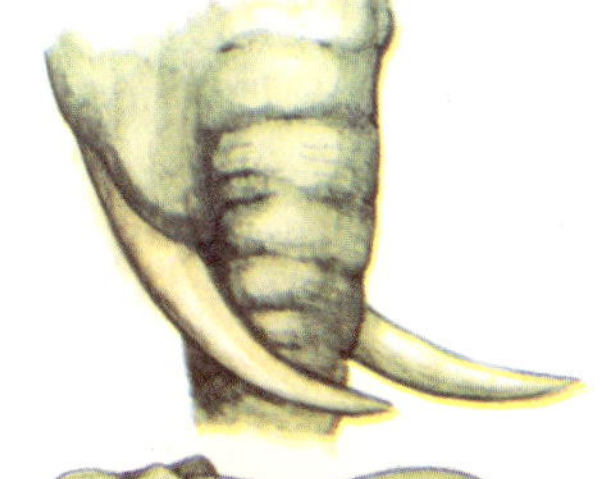

mother
(page 3)

baby
(page 3)

hair
(page 4)

break
(page 7)

stampede
(page 7)

fight
(page 8)

Asia
(page 10)

pin
(page 13)

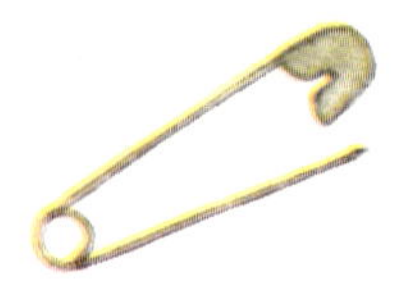

ride
(page 10)

blow
(page 14)

tap
(page 11)

lie down
(page 15)

log
(page 12)

swim
(page 15)

trunk
(page 12)

zoo
(page 16)

circus
(page 17)

trick
(page 17)

hunt
(page 18)

ivory
(page 19)

soldier
(page 20)

India
(page 21)

war
(page 21)

mammoth
(page 22)